PERENNIAL MATH

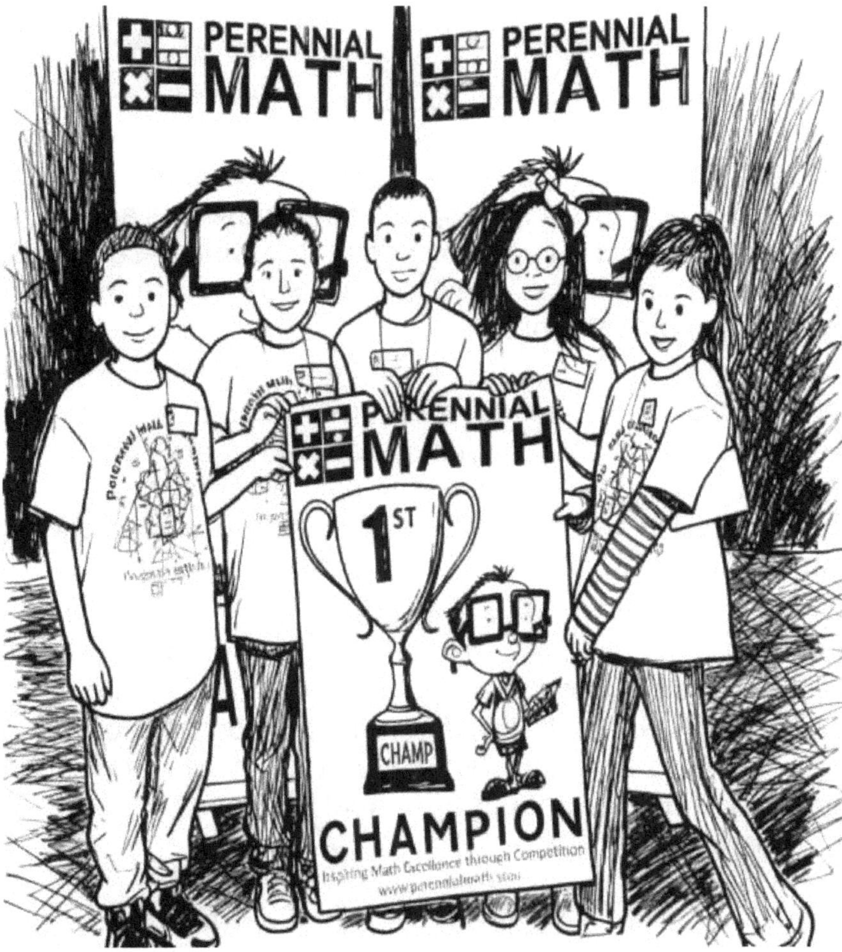

3rd Grade Math Word Problems

About Perennial Math

Perennial Math is a nationally recognized mathematics competition program designed to inspire students to think critically, solve challenging problems, and develop a love for mathematics. Through engaging competitions and tournaments, students strengthen their problem-solving skills, build confidence, and experience the excitement of academic achievement.

Serving students in grades 3–12, Perennial Math offers online competitions, virtual tournaments, and live events that encourage both individual growth and teamwork. Participants have opportunities to earn awards, gain recognition, and compete with students from across the country.

Perennial Math is committed to making mathematics meaningful, accessible, and fun while helping students develop the skills they need for future academic success. Whether in the classroom, at home, or in competition, Perennial Math empowers young learners to become confident and capable mathematical thinkers.

Inspiring Math Excellence through Competition

3rd Grade Math Word Problem Workbook
Table of Contents

CATCH THE MATH BUG

Addition - Adding Within 100

Directions: Read each problem. Show your work. Circle your answer.

1. Mia has 24 crayons. Her teacher gives her 35 more crayons. How many crayons does Mia have now?

2. There are 46 students on the playground. Later, 18 more students come outside. How many students are on the playground now?

3. A farmer picks 32 apples in the morning and 27 apples in the afternoon. How many apples does the farmer pick in all?

4. Ben read 15 pages on Monday and 29 pages on Tuesday. How many pages did Ben read altogether?

5. Mason read 28 pages on Monday and 37 pages on Tuesday. How many pages did Mason read altogether?

Draw It!

Draw pictures or use tally marks to help solve the problem.

Addition - Adding With Regrouping

Directions: Solve each problem. Show your work.

1. A store sold 58 pencils in the morning and 47 pencils in the afternoon. How many pencils were sold in all?

2. A school collected 64 cans of food on Monday and 39 cans on Tuesday. How many cans were collected altogether?

3. Lily has 76 beads. She buys 28 more beads. How many beads does Lily have now?

4. There are 85 birds in one tree and 26 birds in another tree. How many birds are there in all?

5. A library has 134 fiction books and 126 nonfiction books. How many books are there in total?

Illustration Prompt:
Draw two groups and show how you regroup to add.

Addition - Adding Three Numbers

Directions: Add all the numbers to solve the problem.

1. Jack collected 12 shells at the beach, 25 shells near the rocks, and 18 shells by the water. How many shells did Jack collect in all?

2. A bakery sold 24 muffins in the morning, 36 muffins in the afternoon, and 19 muffins in the evening. How many muffins were sold altogether?

3. Three buses brought 45, 38, and 27 students to the zoo. How many students came to the zoo?

4. Emma saved 15 dollars in January, 22 dollars in February, and 13 dollars in March. How much money did Emma save in all?

5. Ethan scored 16 points in the first game, 19 points in the second game, and 22 points in the third game. How many points did Ethan score in all?

Strategy Tip:
Add two numbers first, then add the third number.

Addition - Addition Problems

Directions: Read carefully and solve.

1. A soccer team scored 28 points in the first game and 34 points in the second game. How many points did they score in all?

2. There are 67 blue balloons and 29 red balloons at a party. How many balloons are there altogether?

3. A teacher has 142 math books and 158 reading books. How many books does the teacher have in total?

4. A pet store sold 48 fish on Saturday and 56 fish on Sunday. How many fish were sold over the weekend?

5. A family drove 125 miles on Friday and 187 miles on Saturday. How many miles did they drive in all?

🎨 **Illustration Prompt:**
Draw pictures, number lines, or boxes to solve the problem.

Subtraction - Within 100

Directions: Read each problem. Show your work. Circle your answer.

1. Noah had 68 marbles. He gave 24 marbles to his friend. How many marbles does Noah have now?

2. There were 75 birds sitting on a fence. 38 birds flew away. How many birds are left?

3. Ava baked 54 cookies. Her family ate 27 cookies. How many cookies are left?

4. A toy store had 90 toy cars. They sold 46 toy cars. How many toy cars are left?

5. A farmer had 97 eggs. He sold 54 eggs. How many eggs are left?

Draw It!
Draw pictures or use a number line to help solve the problem.

Subtraction - With Regrouping

Directions: Solve each problem. Show your work.

1. A school had 83 students on the playground. 47 students went back inside. How many students are still on the playground?

2. Liam had 92 trading cards. He gave 58 cards to his cousin. How many cards does Liam have now?

3. A library had 106 books. 49 books were checked out. How many books are left in the library?

4. There were 71 balloons at a party. 36 balloons popped. How many balloons are left?

5. A farmer picked 120 oranges. He sold 68 oranges. How many oranges does the farmer have left?

Illustration Prompt:
Draw base-ten blocks to show how you regroup

Subtraction - Comparing Differences

Directions: Find how many more or how many fewer.

1. Sam has 56 stickers. Ella has 39 stickers. How many more stickers does Sam have than Ella?

2. A red team scored 84 points. A blue team scored 67 points. How many fewer points did the blue team score?

3. There are 95 boys and 78 girls in a school. How many more boys than girls are there?

4. A bookstore sold 132 fiction books and 95 nonfiction books. How many more fiction books were sold?

5. A classroom had 82 pencils. Students used 27 pencils. How many pencils are left?

Strategy Tip:
Subtract the smaller number from the larger number.

Subtraction - Real-Life Problems

Directions: Read carefully and solve.

1. A bus started with 60 passengers. At the first stop, 23 passengers got off. How many passengers are still on the bus?

2. There were 145 apples in a crate. After some were used, 89 apples remained. How many apples were used?

3. A video game gives 100 points to start. After losing points, a player has 64 points. How many points were lost?

4. A class raised $200 for a trip. They spent $78 on supplies. How much money is left?

5. A runner planned to run 150 minutes in a week. She has already run 92 minutes. How many more minutes does she need to run?

Illustration Prompt:
Draw a bar model or number line to help solve the problem.

Multiplication - Equal Groups

Directions: Read each problem. Show your work.

1. There are 4 bags of apples. Each bag has 6 apples. How many apples are there in all?

2. A teacher puts 5 pencils in each box. There are 7 boxes. How many pencils are there in total?

3. A spider has 8 legs. How many legs do 3 spiders have?

4. There are 9 cookies on each plate. If there are 4 plates, how many cookies are there?

5. There are 5 toy cars on each shelf. If there are 8 shelves, how many toy cars are there?

Draw It!

Draw equal groups to help solve the problem.

Multiplication - Repeated Addition

Directions: Use repeated addition or multiplication to solve.

1. A frog jumps 5 feet each time. How far does it jump in 6 jumps?

2. There are 8 crayons in a box. How many crayons are in 5 boxes?

3. A farmer plants 10 seeds in each row. There are 4 rows. How many seeds are planted?

4. A classroom has 6 tables. Each table seats 4 students. How many students can sit at the tables?

5. A bakery puts 3 muffins in each bag. There are 9 bags. How many muffins are there?

Illustration Prompt:
Draw and label the repeated addition.

Multiplication - Arrays

Directions: Use arrays to solve the problems.

1. Chairs are set up in 5 rows with 6 chairs in each row.
How many chairs are there?

2. A garden has 4 rows of flowers. Each row has 8 flowers.
How many flowers are there in all?

3. There are 7 shelves. Each shelf holds 5 books.
How many books are on the shelves?

4. A game board has 9 rows with 3 spaces in each row.
How many spaces are on the board?

5. A farmer plants 10 rows of corn with 8 plants in each row.
How many corn plants are there?

Strategy Tip:
Rows × Columns = Total.

Multiplication - Multiplying by 2, 5, and 10

Directions: Solve each problem.

1. There are 2 wheels on each bike. How many wheels are on 9 bikes?

2. A child earns 5 points each round. How many points after 8 rounds?

3. A teacher gives 10 stickers to each student. There are 6 students. How many stickers are given?

4. A baker makes 5 trays of cookies. Each tray has 10 cookies. How many cookies are made?

5. There are 2 gloves in a pair. How many gloves are in 12 pairs?

Illustration Prompt:
Use skip counting to show your work.

Multiplication - Word Problems with Pictures in Mind

Directions: Read carefully and solve.

1. A pet store has 6 tanks. Each tank has 7 fish. How many fish are in the store?

2. A classroom has 8 rows of desks. There are 4 desks in each row. How many desks are there?

3. A box holds 9 toy cars. How many toy cars are in 5 boxes?

4. A playground has 3 swings. Each swing holds 2 children. How many children can swing at one time?

5. A parking lot has 7 rows. Each row has 5 cars. How many cars are parked altogether?

Draw It!

Draw pictures or arrays to match the problem.

Multiplication - Real-Life Problems

Directions: Solve each problem.

1. A movie theater has 7 rows of seats. Each row has 8 seats. How many seats are in the theater?

2. A pack has 6 juice boxes. How many juice boxes are in 9 packs?

3. A teacher buys 4 packs of markers. Each pack has 12 markers.
How many markers are there?

4. A calendar shows 5 weeks. How many days are there in 5 weeks?

5. A music class has 10 groups. Each group has 3 students. How many students are in the music class?

Think About It:

Which multiplication fact helps you solve fastest?

Division - Sharing Equally

Directions: Read each problem. Show your work.

1. There are 24 cookies to share equally among 6 children. How many cookies does each child get?

2. A teacher has 20 pencils and gives them equally to 5 students. How many pencils does each student get?

3. There are 18 apples placed into 3 baskets. How many apples are in each basket?

4. A bag has 16 marbles that are shared equally into 4 cups. How many marbles go in each cup?

5. A bag has 20 marbles that are shared equally into 5 cups. How many marbles go in each cup?

Draw It!
Draw circles to show how the items are shared.

Division - Grouping

Directions: Find how many groups are made.

1. There are 30 crayons. If 5 crayons go in each box, how many boxes are needed?

2. A baker has 28 muffins. He puts 7 muffins in each tray. How many trays are used?

3. A farmer picks 24 eggs and packs them into cartons of How many cartons are filled?

4. There are 45 students going on a trip. They ride in vans that hold 9 students each. How many vans are needed?

5. A class has 32 markers. They are placed into packs of 8. How many packs are made?

Illustration Prompt:
Draw groups to show how the items are divided.

Division - Using Arrays

Directions: Use arrays to help solve.

1. There are 21 chairs arranged in 3 equal rows. How many chairs are in each row?

2. A garden has 24 flowers planted in 6 equal rows. How many flowers are in each row?

3. A game has 16 players split into 4 equal teams. How many players are on each team?

4. There are 35 stickers placed into 5 equal rows. How many stickers are in each row?

5. There are 24 apples placed into 4 equal baskets. How many apples are in each basket?

Strategy Tip:
Think: total ÷ number of rows = number in each row.

Division - Relating Division to Multiplication

Directions: Use multiplication facts you know to help divide.

1. 32 ÷ 4 = ___
A box holds 32 balls. They are split into 4 groups.

2. 45 ÷ 5 = ___
A teacher has 45 papers and puts them into 5 piles.

3. 36 ÷ 6 = ___
There are 36 cookies shared equally among 6 friends.

4. 40 ÷ 8 = ___
There are 40 crayons placed into boxes of 8.

5. 27 ÷ 3 = ___
A classroom has 27 students split into 3 groups.

Illustration Prompt:
Write the matching multiplication fact.

Division - With Remainders

Directions: Solve and write the remainder if there is one.

1. There are 17 balloons shared equally among 4 children. How many balloons does each child get? How many are left over?

2. A baker has 25 cookies and puts 6 cookies in each bag. How many full bags are made? How many cookies are left?

3. There are 29 students and 5 equal teams.
How many students are on each team? How many are left over?

4. A box holds 10 toys. There are 43 toys.
How many full boxes can be filled? How many toys are left?

5. A crate holds 8 apples. There are 57 apples.
How many full crates can be filled? How many apples are left?

Draw It!
Circle the leftovers.

Division - Real-Life Problems

Directions: Read carefully and solve.

1. A coach has 48 water bottles to give to 6 teams.
How many bottles does each team get?

2. A teacher has 63 stickers and gives 7 stickers to each student. How many students receive stickers?

3. There are 54 apples packed into bags of 9.
How many bags are filled?

4. A library has 72 books placed equally on 8 shelves.
How many books are on each shelf?

5. A pizza shop cuts 36 slices into boxes of 6 slices each.
How many boxes are needed?

Think About It:
Which multiplication fact helped you divide?

Mixed Operations - Choose the Operation

Directions: Read each problem. Decide which operation to use. Show your work.

1. There are 6 boxes of markers. Each box has 8 markers. How many markers are there in all?

2. Mia has 47 stickers. She gives 19 stickers to her friend. How many stickers does Mia have left?

3. A teacher buys 25 notebooks and then buys 18 more. How many notebooks does the teacher have now?

4. There are 32 students in a class. They are put into groups of 4. How many groups are there?

5. There are 30 apples. They are placed into baskets of 5. How many baskets are there?

Think First:
Circle key words like *total*, *left*, *each*, or *groups*.

Mixed Operations - One-Step Word Problems

Directions: Solve each problem. Show your work.

1. A pet store has 9 cages. Each cage holds 5 birds. How many birds are in the store?

2. A library had 84 books. 27 books were checked out. How many books are left?

3. A class collected 6 bags of cans. Each bag has 10 cans. How many cans were collected?

4. Emma had 63 beads. She used 28 beads to make a necklace. How many beads are left?

5. There are 48 cupcakes shared equally among 6 children. How many cupcakes does each child get?

✏️ **Illustration Prompt:**
Draw a picture to solve the problem.

Mixed Operations - Two-Step Problems

Directions: Solve step by step.

1. A baker makes 4 trays of muffins. Each tray has 6 muffins. She sells 10 muffins. How many muffins are left?

2. There are 5 boxes of pencils. Each box has 12 pencils. The teacher gives away 18 pencils. How many pencils remain?

3. A soccer team scores 18 points in the first game and 22 points in the second game. Each goal is worth 2 points. How many goals were scored in all?

4. A school has 3 buses. Each bus carries 24 students. After lunch, 15 students leave early. How many students are still at school?

5. A store receives 8 boxes of pencils, each box containing 10 pencils. If 12 pencils are used, how many are left?

Strategy Tip:
Solve one step at a time.

Mixed Operations - Real-Life Problems

Directions: Read carefully and solve.

1. A farmer picks 36 apples. He puts them into bags of 6 apples each. How many bags are made?

2. There are 45 balloons. 18 balloons pop. How many balloons are left?

3. A toy store has 7 shelves. Each shelf holds 9 toys. How many toys are on the shelves?

4. A class raises $120 for a trip. They spend $45 on supplies. How much money is left?

5. A teacher has 48 stickers. She gives 8 stickers to each student. How many students get stickers?

🎨 **Illustration Prompt:**
Use groups, arrays, or number lines.

Mixed Operations - Word Clues Practice

Directions: Use word clues to choose the correct operation.

1. Each child gets 4 crayons. There are 7 children. How many crayons are needed?

2. A jar holds 50 candies. Some candies are eaten, and now there are 23 left. How many candies were eaten?

3. There are 8 tables with 6 chairs at each table. How many chairs are there in all?

4. A coach has 72 water bottles and gives them equally to 9 teams. How many bottles does each team get?

5. A teacher has 56 pencils and divides them equally among 7 students. How many pencils does each student get?

Think About It:

What word clue helped you decide?

Mixed Operations - Challenge Mixed Problems

Directions: Solve carefully. Show all your work.

1. A store sells 5 packs of stickers. Each pack has 12 stickers. After selling 18 stickers, how many stickers remain?

2. A class has 36 students. They are placed into teams of 6. Each team gets 4 balls. How many balls are needed in total?

3. A baker bakes 8 trays of cookies. Each tray has 10 cookies. He gives 24 cookies to helpers. How many cookies are left?

4. A zoo has 6 groups of animals. Each group has 7 animals. Some animals are moved, and now there are 29 animals left. How many animals were moved?

5. A park has 4 sections with 9 ducks in each section. Now there are 20 ducks left. How many ducks were moved?

Think It Through:
Underline the important information.

Money - Counting Coins

Directions: Count the money. Show your work.

1. Sam has 3 quarters and 2 dimes. How much money does Sam have?

2. Lily has 4 dimes and 5 pennies. How much money does Lily have?

3. A jar has 2 nickels and 6 pennies. How much money is in the jar?

4. Ben has 1 quarter, 3 nickels, and 2 pennies. How much money does Ben have?

5. Emma has 1 quarter, 2 dimes, and 5 pennies. How much money does she have?

Draw It!

Draw each coin and label its value.

Money - Adding Money

Directions: Add the money amounts.

1. A toy costs $0.45. A ball costs $0.32. How much do they cost together?

2. Mia buys a notebook for $1.25 and a pencil for $0.40. How much does she spend in all?

3. A book costs $2.75. A bookmark costs $0.85. What is the total cost?

4. Jake buys a snack for $1.60 and a drink for $1.25. How much does he spend?

5. A game costs $3.45. A poster costs $2.30. How much do they cost together?

✏️ **Illustration Prompt:**
Draw coins or bills to help solve the problem.

Money - Subtracting Money

Directions: Find how much money is left.

1. Emma has $5.00. She buys a toy for $2.35.
How much money does Emma have left?

2. A snack costs $1.75. You pay with $5.00. How much change do you get?

3. A book costs $6.40. You pay with $10.00. How much change do you get?

4. Noah has $4.50. He spends $1.85. How much money is left?

5. Noah has $12.30. He spends $4.90. How much is left?

Strategy Tip:
Line up the decimals.

Money - Making Change

Directions: Solve each problem.

1. A pencil costs $0.65. You pay with $1.00. How much change do you get?

2. A sandwich costs $3.20. You pay with $5.00. How much change do you get?

3. A toy costs $7.45. You pay with $10.00. How much change do you get?

4. A book costs $4.80. You pay with $10.00. How much change do you get?

5. A drink costs $1.95. You pay with $5.00. How much change do you get?

Illustration Prompt:
Draw bills and coins for the change.

Money - Real-Life Problems

Directions: Read carefully and solve.

1. A student saves $2.00 each week. How much money will the student save in 6 weeks?

2. A teacher buys 4 notebooks at $1.50 each. How much does the teacher spend?

3. A movie ticket costs $6.00. How much do 5 tickets cost?

4. A baker sells cupcakes for $2.00 each. How much money does the baker earn by selling 9 cupcakes?

5. A child has $20.00 and buys a game for $14.75. How much money is left?

Think About It:
Should you add, subtract, or multiply?

Money - Challenge Problems

Directions: Solve step by step.

1. A toy costs \$3.50. A book costs \$4.25. How much do they cost together?

2. You buy 3 snacks at \$1.25 each. How much do you spend?

3. You have \$10.00. You spend \$6.80. How much money is left?

4. A class earns \$5.00 each day for 8 days. How much money does the class earn?

5.A group collects \$7.00 each day for 6 days. How much money do they collect?

Show Your Work:
Use pictures or equations.

Measurement - Length (Inches and Feet)

Directions: Solve the problems. Show your work.

1. A pencil is 7 inches long. A pen is 5 inches long. How much longer is the pencil than the pen?

2. A rope is 12 feet long. A piece 5 feet is cut off. How many feet of rope are left?

3. A table is 36 inches long. A chair is 18 inches long. How much longer is the table?

4. A garden is 24 feet long. Another garden is 16 feet long. What is the total length of both gardens?

5. One sidewalk is 42 feet long. Another sidewalk is 18 feet long. What is the total length of both sidewalks?

Draw It!

Use a ruler or draw line segments to show lengths.

Measurement - Weight (Pounds and Ounces)

Directions: Solve the problems.

1. A watermelon weighs 8 pounds. A cantaloupe weighs 3 pounds. How much heavier is the watermelon?

2. A bag of apples weighs 12 pounds. A bag of oranges weighs 7 pounds. What is the total weight?

3. A pumpkin weighs 15 pounds. 6 pounds are cut off. How much pumpkin is left?

4. A sack of flour weighs 10 pounds. Another sack weighs 14 pounds. What is the combined weight?

5. A crate of vegetables weighs 20 pounds. Another crate weighs 8 pounds. What is the total weight?

✏️ **Illustration Prompt:**
Draw scales or stacks to compare weights.

Measurement - Capacity (Liters and Cups)

Directions: Solve the problems.

1. A jug holds 8 cups of water. A pitcher holds 5 cups. How many cups can both hold together?

2. A bottle holds 12 liters of juice. 4 liters are poured out. How many liters remain?

3. A cup holds 3 cups of milk. How many cups are in 4 cups of milk?

4. A tank holds 15 liters. Another tank holds 9 liters. What is the total capacity?

5. A container holds 14 liters. Another container holds 6 liters. What is the total capacity?

Draw It!
Use labeled containers to show amounts.

Measurement - Time (Hours and Minutes)

Directions: Solve the problems.

1. School starts at 8:00 AM and ends at 3:00 PM.
How many hours is school?

2. A movie starts at 2:15 PM and ends at 4:00 PM.
How many minutes long is the movie?

3. Liam practiced piano for 30 minutes in the morning and 45 minutes in the evening.
How many minutes did he practice in all?

4. A train leaves at 10:00 AM and arrives at 1:30 PM.
How many hours and minutes was the trip?

5. A bus leaves at 9:20 AM and arrives at 1:05 PM.
How many hours and minutes was the trip?

✏️ **Illustration Prompt:**
Draw clocks to show the time.

Measurement - Mixed Problems

Directions: Solve step by step.

1. A garden is 18 feet long and 12 feet wide. What is the total length if you walk all the sides *(perimeter)*?

2. A container holds 20 liters of water. 7 liters are poured out. How many liters remain?

3. A rope 15 feet long is cut into 3 equal pieces. How long is each piece?

4. A bag of sugar weighs 12 pounds. Another bag weighs 9 pounds. How much do both weigh together?

5. A box of apples weighs 16 pounds. Another box weighs 12 pounds. How much do they weigh together?

Strategy Tip:
Line up the units before calculating.

Measurement - Real-Life Problems

Directions: Read carefully and solve.

1. A bookshelf is 36 inches tall. A desk is 30 inches tall. How much taller is the bookshelf?

2. A water tank holds 50 liters. 18 liters are used. How many liters are left?

3. A jogging track is 400 meters. If Sarah runs 3 laps. How many meters does she run?

4. A bag of flour weighs 10 pounds. A bag of sugar weighs 8 pounds. What is the total weight?

5. A recipe needs 2 cups of milk. You double the recipe. How many cups of milk do you need?

Illustration Prompt:
Draw a diagram for at least one problem.

Fractions - Understanding Fractions

Directions: Read and solve. Draw a picture if needed.

1. Mia ate 1 out of 4 slices of pizza.
What fraction of the pizza did Mia eat?

2. A chocolate bar is divided into 8 equal pieces.
Liam eats 3 pieces. What fraction did he eat?

3. There are 10 apples. 5 are red and the rest are green.
What fraction of the apples are red?

4. A pie is cut into 6 equal slices. Emma eats 2 slices.
What fraction of the pie is left?

5. A sandwich is cut into 8 equal pieces. Mia eats 2 pieces.
What fraction of the sandwich remains?

Draw It!
Color the parts to show the fraction.

Fractions - Fractions of a Set

Directions: Solve using fractions of a group.

1. There are 12 candies. 1/3 of them are chocolate. How many candies are chocolate?

2. A basket has 18 eggs. 1/2 of them are brown. How many brown eggs are there?

3. A classroom has 24 students. 1/4 of the students are absent. How many students are absent?

4. A pack has 20 pencils. 3/5 of them are yellow. How many pencils are yellow?

5. A box has 25 crayons. 2/5 of them are red. How many are red?

✏️ **Illustration Prompt:**
Draw circles or squares to divide into fractions.

Fractions - Adding Fractions (Same Denominator)

Directions: Solve the problems.

1. Mia ate 1/6 of a cake in the morning and 2/6 in the evening. How much of the cake did she eat in total?

2. Liam drank 3/8 of a bottle of juice. Later he drank 2/8 more. How much juice did he drink altogether?

3. A pizza has 8 slices. Ben ate 3/8 and Emma ate 4/8. What fraction of the pizza was eaten?

4. A ribbon is 12 inches. 5/12 is red, 4/12 is blue. What fraction of the ribbon is red or blue?

5. A rope is 15 feet long. 4/15 is green and 7/15 is yellow. What fraction of the rope is green or yellow?

Draw It!

Use shaded parts to add the fractions.

Fractions - Subtracting Fractions (Same Denominator)

Directions: Solve carefully.

1. A chocolate bar has 6 pieces. Liam eats 4/6. How many pieces are left?

2. Mia drank 5/8 of juice. Later she pours out 2/8. How much juice is left?

3. A pie is cut into 10 slices. 7/10 are eaten. How much of the pie remains?

4. A ribbon is 9 inches long. 6/9 is used. How much of the ribbon is left?

5. A piece of yarn is 24 inches long. 8/24 is used. How much of the yarn is left?

Illustration Prompt:
Shade what is left.

Fractions - Real-Life Problems

Directions: Solve step by step.

1. A cake is cut into 8 slices. 3 slices are chocolate and the rest are vanilla. What fraction is vanilla?

2. There are 20 marbles. 3/4 are blue. How many marbles are blue?

3. A garden has 12 flowers. 2/3 are tulips. How many tulips are there?

4. Liam read 5/10 of a book yesterday and 3/10 today. How much of the book has he read in total?

5. A pizza is cut into 12 slices. 7 slices are eaten. What fraction remains?

Illustration Prompt:
Draw the pizza, cake, or marbles to visualize fractions.

Fractions - Challenge Problems

Directions: Solve carefully.

1. A ribbon is 15 inches long. 2/5 is red, 3/5 is blue. How many inches are red? How many inches are blue?

2. There are 24 cookies. 1/3 are chocolate, 1/4 are vanilla. How many cookies are chocolate and vanilla?

3. A class has 30 students. 2/5 are boys. How many boys are in the class?

4. A pie is divided into 10 slices. Ben ate 3/10 and Emma ate 4/10. How many slices are left?

5. A sandwich is cut into 10 pieces. Ethan ate 2/10 and Chloe ate 5/10. How many pieces are left?

Show Your Work:
Use drawings or number lines to help solve.

Geometry - Identifying Shapes

Directions: Read and solve.

1. A stop sign is shaped like an octagon. How many sides does it have?

2. A picture frame is rectangular. How many sides does a rectangle have?

3. A coin is circular. How many corners does it have?

4. A triangular sandwich is cut into 3 slices. How many sides does each slice have?

5. A rectangular cake is cut into 2 rectangular slices. How many sides does each slice have?

Draw It!
Draw each shape and label its sides.

Geometry - Counting Sides and Angles

Directions: Solve the problems.

1. A square has 4 sides. How many corners does it have?

2. A rectangle has 4 sides. Two sides are longer and two are shorter. How many right angles does it have?

3. A triangle has 3 sides. How many angles does it have?

4. A pentagon has 5 sides. How many corners does it have?

5. An octagon has 8 sides. How many corners does it have?

✏️ **Illustration Prompt:**
Draw shapes and label sides and angles.

Geometry - Perimeter of Shapes

Directions: Find the perimeter. Add the sides.

1. A square has sides of 6 cm. What is the perimeter?

2. A rectangle is 5 cm long and 3 cm wide. What is the perimeter?

3. A triangle has sides of 4 cm, 5 cm, and 6 cm. What is the perimeter?

4. A pentagon has sides of 3 cm each. What is the perimeter?

5. A pentagon has sides of 7 cm each. What is the perimeter?

Draw It!

Label each side with its length.

Geometry - Area of Rectangles

Directions: Solve the problems. Use **length × width**.

1. A rectangle is 4 m long and 3 m wide. What is the area?

2. A garden is 6 m long and 5 m wide. What is the area?

3. A classroom rug is 8 ft long and 7 ft wide. What is the area?

4. A window is 5 m long and 2 m wide. What is the area?

5. A carpet is 4 m long and 5 m wide. What is the area?

Illustration Prompt:
Draw the rectangle and shade the area.

Geometry - 3D Shapes

Directions: Identify and solve problems about 3D shapes.

1. A cube has 6 faces. How many edges does it have?

2. A rectangular prism has 12 edges. How many faces does it have?

3. A cylinder has 2 circular faces. How many curved surfaces does it have?

4. A cone has 1 circular face. How many edges does it have?

5. A triangular prism has 5 faces. How many edges does it have?

Draw It!

Sketch each 3D shape and label faces, edges, and vertices.

Geometry - Real-Life Problems

Directions: Solve step by step.

1. A garden is shaped like a rectangle, 8 m long and 3 m wide. What is the perimeter and area?

2. A picture frame is 10 inches long and 7 inches wide. What is the area?

3. A cube-shaped box has sides of 4 cm. How many faces does it have? How many edges?

4. A triangle-shaped park has sides of 5 m, 6 m, and 7 m. What is the perimeter?

5. A rectangular playground is 20 m long and 10 m wide. What is the area?

Illustration Prompt:
Draw each shape and label dimensions.

Mixed Review - Addition & Subtraction

Directions: Solve each problem.

1. Mia has 45 stickers. She gives 18 to her friend. How many stickers does Mia have left?

2. A farmer collects 36 eggs and buys 24 more. How many eggs does he have in total?

3. A class has 32 pencils. 14 are used. How many are left?

4. There are 27 apples. 9 apples are eaten. How many remain?

5. There are 30 toy cars in a box. 13 are taken out. How many are left?

Draw It!
Use a number line or pictures to show subtraction or addition.

Mixed Review - Multiplication & Division

Directions: Solve each problem.

1. There are 7 boxes of crayons. Each box has 6 crayons. How many crayons are there?

2. A cake is cut into 12 slices. 4 children share the cake equally. How many slices does each child get?

3. A soccer team has 11 players. If there are 3 teams, how many players are there in total?

4. A baker makes 24 cookies. She puts 6 cookies in each bag. How many bags are filled?

5. A pet store has 32 fish. They are placed into tanks with 4 fish in each tank. How many tanks are needed?

Illustration Prompt:
Draw equal groups or arrays.

Mixed Review - Money & Time

Directions: Solve each problem.

1. A toy costs $3.25. You pay with $5.00. How much change do you get?

2. Mia saves $2.00 each week. How much will she save in 5 weeks?

3. School starts at 8:00 AM and ends at 2:30 PM. How long is school?

4. A snack costs $1.50 and a drink costs $0.75. How much do both cost together?

5. A notebook costs $2.35 and a pen costs $1.15. How much do both cost together?

Illustration Prompt:
Draw coins or a clock to show your work.

Mixed Review - Fractions & Measurement

Directions: Solve each problem.

1. A pizza is cut into 8 slices. Liam eats 3 slices. What fraction did he eat?

2. A ribbon is 12 inches long. 5 inches are red. How many inches are not red?

3. A rectangle is 6 m long and 4 m wide. What is the area?

4. A chocolate bar has 10 pieces. 7 are eaten. What fraction remains?

5. A garden is 20 feet long and 10 feet wide. What is the perimeter?

Draw It!
Use diagrams or fraction bars.

Mixed Review - Geometry

Directions: Solve each problem.

1. A cube has 6 faces. How many edges does it have?

2. A triangle has sides of 5 cm, 6 cm, and 7 cm. What is the perimeter?

3. A rectangle is 8 m long and 3 m wide. What is the area?

4. A circle has no corners. How many sides does it have?

5. A rectangular prism has 6 faces and 12 edges. How many vertices does it have?

Illustration Prompt:
Draw each shape and label sides, edges, or faces.

Mixed Review - Challenge Problems

Directions: Solve step by step.

1. A class collects $12 on Monday, $15 on Tuesday, and $18 on Wednesday. How much money do they collect in total?

2. A baker makes 5 trays of cupcakes. Each tray has 8 cupcakes. He sells 18 cupcakes. How many cupcakes are left?

3. A pizza has 12 slices. 5 slices are pepperoni. 3 slices are cheese. What fraction is not pepperoni or cheese?

4. A playground is 20 m long and 10 m wide. What is the perimeter and area?

5. A box holds 36 pencils. 6 pencils go in each pack. How many packs are needed?

Show Your Work:
Use drawings, number lines, or equations.

ANSWER KEY

Page 1
1.59
2.64
3.59
4.44
5.65

Page 2
1.105
2.103
3.104
4.111
5.260

Page 3
1.55
2.79
3.110
4.50
5.57

Page 4
1.62
2.96
3.300
4.104
5.312

Page 5
1.44
2.37
3.27
4.44
5.43

Page 6
1.36
2.34
3.57
4.35
5.52

Page 7
1.17
2.17
3.17
4.37
5.55

Page 8
1.37
2.56
3.36
4.122
5.58

Page 9
1.24
2.35
3.24
4.36
5.40

Page10
1.30
2.40
3.40
4.24
5.27

Page11
1.30
2.32
3.35
4.27
5.80

Page12
1.18
2.40
3.60
4.50
5.24

Page 13
1.42
2.32
3.45
4.6
5.35

Page 14
1.56
2.54
3.48
4.35
5.30

Page 15
1.4
2.4
3.6
4.4
5.4

Page 16
1.6
2.4
3.4
4.5
5.4

Page 17
1.7
2.4
3.4
4.7
5.6

Page 18
1.8
2.9
3.6
4.5
5.9

Page 19
1.4 R1
2.4 R1
3.5 R4
4.4 R3
5.7 R1

Page 20
1.8
2.9
3.6
4.9
5.6

Page 21
1.48
2.28
3.43
4.8
5.6

Page 22
1.45
2.57
3.60
4.35
5.8

Page 23
1.14
2.42
3.20
4.57
5.68

Page 24
1.6
2.27
3.63
4.75
5.6

Page 25
1.28
2.27
3.48
4.8
5.8

Page 26
1.42
2.24
3.56
4.13
5.16

Page 27
1.$0.95
2.$0.45
3.$0.16
4.$0.42
5.$0.50

Page 28
1.$0.77
2.$1.65
3.$3.60
4.$2.85
5.$5.75

Page 29
1.$2.65
2.$3.25
3.$3.60
4.$2.65
5.$7.40

Page 30
1.$0.35
2.$1.80
3.$2.55
4.$5.20
5.$3.05

Page 31
1.$12.00
2.$6.00
3.$30.00
4.$18.00
5.$5.25

Page 32
1.$7.75
2.$3.75
3.$3.20
4.$40.00
5.$42.00

Page 33
1.2 inches
2.7 feet
3.18 inches
4.40 feet
5.60 feet

Page 34
1.5 pounds
2.19 pounds
3.9 pounds
4.24 pounds
5.28 pounds

Page 35
1.13 cups
2.8 liters
3.12 cups
4.24 liters
5.20 liters

Page 36
1.7 hours
2.105 minutes
3.75 minutes
4.3 hours 30 minutes
5.3 hours 45 minutes

Page 37
1.60 feet
2.13 liters
3.5 feet
4.21 pounds
5.28 pounds

Page 38
1.6 inches
2.32 liters
3.1,200 meters
4.18 pounds
5.4 cups

Page 39
1.1/4
2.3/8
3.5/10 or 1/2
4.4/6 or 2/3
5.6/8 or 3/4
Page 40
1.4 candies
2.9 eggs
3.6 students
4.12 pencils
5.10 red crayons
Page 41
1.3/6 or 1/2
2.5/8
3.7/8
4.9/12 or 3/4
5.11/15
Page 42
1.2/6 or 1/3
2.3/8
3.3/10
4.3/9 or 1/3
5.16 inches
Page 43
1.5/8
2.15 marbles
3.8 tulips
4.8/10 or 4/5
5.5/12
Page 44
1.Red: 6 inches, Blue: 9 inches
2.Chocolate: 8 cookies, Vanilla: 6 cookies
3.12 boys
4.3 slices left
5.3 pieces left
Page 45
1.8 sides
2.4 sides
3.0 corners
4.3 sides
5.4 sides
Page 46
1.4 corners
2.4 right angles
3.3 angles
4.5 corners
5.8 corners
Page 47
1.24 cm
2.16 cm
3.15 cm
4.15 cm
5.35 cm
Page 48
1.12 m^2
2.30 m^2
3.56 ft^2
4.10 m^2
5.20 m^2
Page 49
1.12 edges
2.6 faces
3.1 curved surface
4.1 edge
5.9 edges
Page 50
1.Perimeter: 22 m, Area: 24 m^2
2.Area: 70 in^2
3.Faces: 6, Edges: 12
4.Perimeter: 18 m
5.Area: 200 m^2
Page 51
1.27 stickers
2.60 eggs
3.18 pencils
4.18 apples
5.17 toy cars
Page 52
1.42 crayons
2.3 slices each
3.33 players
4.4 bags
5.8 tanks
Page 53
1.$1.75 change
2.$10.00
3.6 hours 30 minutes
4.$2.25
5.$3.50
Page 54
1.3/8
2.7 inches
3.24 m^2
4.3/10
5.60 feet
Page 55
1.12 edges
2.18 cm
3.24 m^2
4.0 sides
5.8 vertices
Page 56
1.$45
2.22 cupcakes
3.4/12 or 1/3
4.Perimeter: 60 m, Area: 200 m^2
5.6 packs

www.ingramcontent.com/pod-product-compliance
Lightning Source LLC
Chambersburg PA
CBHW071400230926
51846CB00002B/607
* 9 7 8 1 9 6 8 4 7 7 1 9 6 *